FOOTPRINT DESIGN MANUAL FOR LOCAL ROADS

SPONSORED BY
The Transportation and Development Institute (T&DI)
of the American Society of Civil Engineers

AUTHORED BY
The Local Roads and Streets Committee of the Transportation and
Development Institute of the American Society of Civil Engineers

Published by the American Society of Civil Engineers

Cataloging-in-Publication Data on file with the Library of Congress.

American Society of Civil Engineers
1801 Alexander Bell Drive
Reston, Virginia, 20191-4400

www.pubs.asce.org

Preface

In response to limited design guidance available for Resurfacing, Restoration, and Rehabilitation (RRR) projects, functionally classified as "Local Roads" of the National Highway System, this committee has specifically undertaken the task to develop systematic methods that relate to Resurfacing, Restoration, and Rehabilitation (RRR) projects. The committee's critical reviews and findings, from publications like Transportation Research Board," *Special Report 214"*, *1987 (1)*; AASHTO, American Association of State Highway and Transportation Officials, *"Guidelines for Geometric Design of Very Low-Volume Local Roads (ADT ≤ 400)"*, *2001 (2)*; AASHTO, Association of State Highway and Transportation Officials *"Geometric Design for Resurfacing, Restoration, and Rehabilitation (RRR) of Streets"*, *1977 (3)*, and many other local agency internal publications were used to make judgments about the relationship between safety and key highway features. For several design features, the committee found sufficient evidence to support quantitative relationships between safety and design improvements. However, the relationships must be reviewed as approximate in nature. Although the relationships are based on the best available data, they could be substantially changed be the results of future research.

The *Code of Federal Register 2007, Title 23: Highways, Part 625 – Design Standards for Highways, § 625.2 Policy (b),* states "Resurfacing, restoration, and rehabilitation (RRR) projects, other than those on the Interstate system and other freeways, shall be constructed in accordance with standards which preserve and extend the service life of highways and enhance highway safety. Resurfacing, restoration, and rehabilitation work includes placement of additional surface material and/or other work necessary to return an existing roadway, including shoulders, bridges, the roadside, and appurtenances to a condition of structural of functional adequacy."

In addition to publications used for (RRR) work, other research and documents were used in the production of this document: AASHTO, Association of State Highway and Transportation Officials, *"Roadside Design Guide" (4)*, National Cooperative Highway Research Program, *"Report 350, Recommended Procedures for the Safety Performance Evaluation of Highway Features"*, Transportation Research Record 1599 (5), *"Guardrail Need: Embankments and Culverts"*, Transportation Research Board, Washington, DC, 1997 (6).

Drawing primarily on case studies of current RRR practices and analyses of safety cost-effectiveness, the committee has recommended practices that encompass the entire RRR process but with special focus on design. The committee's recommendations in this document are intended to serve as guidance. Engineering judgment based of local conditions is paramount in fulfilling the tasks to improve an existing roadway and to improve safety.

Acknowledgements

Special appreciation is expressed to the members of the Local Roads and Streets Committee of the Transportation and Development Institute of ASCE that contributed to the development of this document.

Joe W. Ruffer, P.E., F.ASCE, Co-Author
James D. Foster, Co-Author
Cornelius W. Andres, P.E., A.M.ASCE
Larry W. Emig, P.E., F.ASCE
Andrew E. Ramisch, P.E., M.ASCE
Eugene R. Russell, Ph.D., P.E., F.ASCE
Roger E. Smith, Ph.D., P.E., F.ASCE
John C. Vancor, M.ASCE

Contents

I. Introduction

It has become apparent ever since road construction began that funding available for resurfacing, restoration and rehabilitation (RRR) of local roads and streets will be insufficient to improve existing roadways to the geometric standards desirable for major reconstruction and new construction at a rate equal to that at which pavements are deteriorating. Available funds are expected to remain essentially constant or perhaps even decrease while at the same time construction costs are increasing.

In addition to costs, upgrading highways to guidance levels recommended for new construction (*AASHTO A Policy on Geometric Design of Highways and Streets*), impacts the environment of abutting areas and communities in the vicinity. The social and economic costs to the community must be balanced against improved service to the traveling public. Many publications like *"A Guide for Achieving Flexibility in Highway Design" (7), "Roadside Design Guide" (8), "Guidelines for Geometric Design of Very Low-Volume Local Roads (ADT ≤ 400)" (2)*, have shown that flexibility has to be a part of the design process.

Previously stated reasons provide the background and need for new geometric guidelines for resurfacing, restoration, and rehabilitation (RRR) projects. When the designer determines that RRR design criteria should be used, this manual is intended to provide guidelines to follow in the design process. The design engineer should consider each project individually using engineering judgment to determine what improvements are feasible within available funding to provide a facility that will serve the public at a reasonable level of safety and comfort. This innovative approach to design is essential in order to give designers options to be use the limited roadway resources to meet the pressing needs of improving function and safety characteristics to the extent possible of the roadway systems in a cost effective manner.

This guide has been developed to provide the designer flexibility by presenting minimum values for design and recognizing that engineering judgment should be used to obtain the traffic service and safety benefits possible within existing conditions and constraints. This guide is only applicable to roadways functionally classified as "Local". For higher functionally classified roadways other publications should be used.

The primary purposes of RRR projects are to provide a better riding surface, preserve pavement structural section, increase safety, and to improve operating conditions, to the most feasible degree possible. In addition to the primary objectives, it may be possible in some cases to consider secondary objectives appropriate to a project to an extent that is financially and environmentally acceptable.

The following list of objectives (not in priority order, or all inclusive) may be considered.

Primary Objectives:

- Improve surface smoothness
- Extend service life
- Restore cross-slope
- Improve superelevation
- Improve skid resistance
- Restore deteriorated bridge decks
- Reconstruct sections of pavement structure
- Widen pavement and shoulders
- Flatten front slopes
- Improve drainage
- Improve pipe-ends treatments
- Extend culverts
- Upgrading traffic control devices
- Improve sight distance
- Improve site-specific crash locations

Secondary Objectives:

- Increase vertical and horizontal clearance to obstructions
- Intersection improvements and channelization
- Provide paved shoulders
- Provide for control of erosion
- Install new types of traffic control devices
- Provide curbing, sidewalks, ADA ramps (only in built-up areas)
- Provide bikeways
- Install street lighting
- Improved landscaping
- Flatten back slopes
- Construct closed drainage systems

The present right-of-way (ROW) may be adequate to accomplish the above improvements. In some cases minor ROW acquisitions or easements may be require. Deficiencies in some existing roadways systems are usually identified by sufficiency ratings, crash data, skid tests, maintenance reports, road safety audits, and in some cases, suggestions from the public.

Often attention to the overall appearance of the roadway, as it is being improved, will result in a product that is more readily accepted by the community. Examples could be the inclusion of curb and sidewalks in urbanized areas, the relocation of utility poles away from the edge of pavement, or the addition of wider shoulders and flattened front slopes. The cost of such improvements must be carefully weighed against benefits available from an equivalent project elsewhere.

II. Types of Projects

Preservation or Maintenance

These are projects where its primary objective is to preserve and extend the service life of existing roads. This is an important activity for the preservation of a roadway. This type of work would typically not have any additional items of work that would upgrade its present condition. This guide would not be applicable for this type of activity.

General RRR

RRR projects are divided into three categories

Resurfacing

While this category is primarily for pavement resurfacing, other types of work may be included such as short sections of pavement reconstruction, jacking concrete slabs, and joint replacement and/or repair. It might also include widening of narrow lanes, shoulders, traffic control devices, channelization work, barriers, and some drainage improvements. Locations, which have proven to be hazardous, should be corrected. Usually no additional rights-of-way are required.

Restoration

This type of work would return road or structures to the condition of original construction. Some intersections may need additional capacity. There could be some need for curbing, sidewalks, channelization, drainage improvements, etc. Resurfacing or pavement reconstruction to improve wet weather safety is included that will enable existing pavement to perform satisfactorily for substantial time periods. New and upgraded traffic control devices are commonly needed. Some additional right-of-way may be necessary. Consideration may be given to improving an isolated grade, curve, or sight distance by construction or traffic control measures.

Rehabilitation

Traffic service improvements and some betterment needs in this category may be of equal or greater importance than the need to improve the riding quality of the pavement. These roads are usually found in urban areas or suburban areas where land use along the facility has intensified over the years. There is a great need to provide continuous through or auxiliary lanes in order to reduce traffic bottlenecks and improve traffic service and safety. Safety should be given close attention with emphasis on features having crash history and those known to have high potential for crashes. Often a closed drainage system may be appropriate. Curbing and sidewalks may be desirable.

Resurfacing of the existing pavement is usually included. In some cases, complete pavement structure replacement or enhancements that extend the service life and/or improve its load carrying capability for specified sections may be called for. Retaining walls may be required. Bridge widening, deck replacement, or railing upgrading may be necessary. New and upgraded traffic control devices are commonly needed. Some additional right-of-way may be necessary. Consideration may be given to improving isolated grade, curve, or sight distance by construction or traffic control measures.

Reconstruction

Work that would increase the functional classification of the roadway, improve the level-of-service (LOS), increase capacity, increase design speed, and/or improve horizontal and vertical alignment along a substantial length of a roadway would be reconstruction activities. This publication is not intended for these types of improvements. Other guidance should be obtained.

III. Establishing Geometric Guidelines

General

RRR projects should apply design criteria that will allow some flexibility in order to adjust to actual field conditions. Therefore, the geometric information in this guide is generally the minimum considered acceptable. It is intended that engineering judgment be exercised to determine where it may be feasible to design above these minimums in order to insure the greatest traffic service and safety improvements possible within existing conditions and constraints.

Traffic Data

The projects covered by this guideline are undertaken primarily to meet specific current needs and are designed to improve a greater portion of the roadway system within funds available. Therefore, the basic thrust of RRR projects must be to satisfy existing traffic conditions. The present level-of-service will be maintained or improved if found to be cost effective.

Current data that should be available during the design is as follows:

1. ADT and/or DHV
2. Crash locations and descriptions
3. Turning movements at major traffic generators
4. Any known "future developments" that could impact the roadway

Because of the varying degree of projects, RRR improvements and costs should be developed on the basis of a 5 or 10-year traffic forecast. If existing volumes are high and conditions are restricted, only minimal increase in capacity may be realized.

Speed

It is common practice when full reconstruction is being considered to relate forecasted traffic volumes to specific design criteria including design speed. Higher forecasted traffic volumes usually require higher guidelines. However, when traffic volumes become moderate to heavy, it is usually because the roadway is approaching or is within an urban area. Thus, the ability to apply higher guidelines becomes increasingly difficult and costly because of adjacent land uses. For this reason, many projects often cannot be implemented when it is necessary to meet the guidelines of new construction.

It is apparent from the above discussion that if existing roadways are to be maintained and improved within strict constraints and minimal social and environmental impact, a more cost-effective approach is essential. The desirable design should accommodate the current running speed but a minimum design speed should not be established.

Advisory speed reduction signs attached to curve signs may be utilized for horizontal curvature as is the present practice and/or any other traffic control devices available, with the same for vertical curves. Transportation Research Board, " *Special Report 214", 1987,* suggests that when the difference between the current running speed and corresponding design speed of a horizontal curve exceeds 15 mph, or vertical curve exceeds 20 mph, additional considerations should be given to corrective work or to provide additional warning devices in order to avoid large changes of running speed.

It is important when considering a RRR project for a section of roadway to consider the geometric conditions beyond the portion to be improved. Every attempt should be made to maintain a uniformly safe running speed for a significant segment of roadway. Considerable consideration should be given to the transition point between portions of a roadway having different design speeds. The greater the change in design speeds the higher the demand that will be placed on driver expectancy. Consistency in design is paramount to driver expectation, one without abrupt changes in section or alignment. If confronted with a transition point, for whatever reason, attempts should be carefully planned to advise the driver well in advance of this change point. The Manual on Uniform Traffic Control Devices (MUTCD) should be referenced and implemented as needed as part of the RRR project.

IV. Design Criteria Recommendations

Significant improvements in safety should be systematically designed into each roadway RRR project. Designers should seek opportunities specific to each project and apply sound safety and traffic engineering principles. Attention to safety, along with documentation of the design process improves design decisions. The design practice should incorporate the following recommendations.

Recommendation 1: Review Current Conditions

Designers should review existing physical and operational conditions affecting safety:

- Conduct and document a thorough site inspection of all physical elements and geometry within the roadway limits.
- Analyze existing roadway users, functional classification, ADT, and average speeds.
- Analyze crash data, to include field inspection, and concerns expressed by the public.
- A combination of different elements may contribute to possible reasons for a crash location.

Recommendation 2: Determine Project Scope

In addition to pavement repairs, the designers should consider, where appropriate, to incorporate; intersection, roadside, and traffic control improvements that may enhance safety. Based on recommendation #1 the designer should:

- Determine site-specific locations where physical elements should be replaced or improved. The designer should field review the roadway for; driveways hidden because of roadway geometry, especially if the driveway is used by large trucks or farm machinery, intersections with limited sight distance, sharp horizontal or vertical curves, narrow bridge, drainage areas close to the pavement, headwalls, obstructions within the right-of-way, etc.

Poorly maintained sign doesn't help the driver's awareness of curve ahead

- Include low-cost improvements, like replacing roadway sign that meet M.U.T.C.D current requirements in a project, can vastly enhance the appearance of a project as well aid the driver's decision making. Signs should command respect of the action being advised to the user.

- Good travelway cross-section. Additional safety benefits include a paved shoulder (reducing pavement edge drop) and gentle sloping frontslope (helps the errant driver to recover back to the travelway). However, the vertical headwall can cause serious injuries when struck.
- Determine site-specific locations where crash data indicates the need for additional improvements. The designer should review crash data information and may develop collision diagrams.
- It is important to know the functional classification of the roadway. Some adjacent elements along the roadside may not be appropriate if an errant vehicle leaves the travelway at high speeds.

- Narrowing of the travelway creates a potential danger for a head-on collision, especially at night.

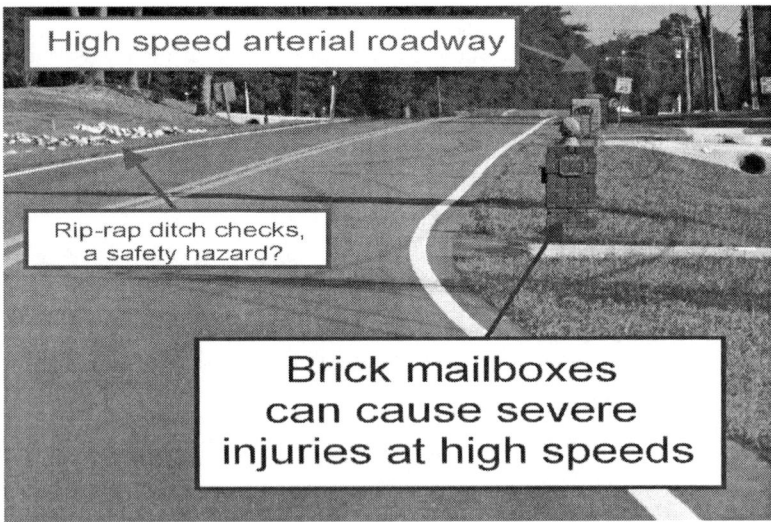

- You must know the proper devices to use and where to use. Will a motorcycle, bicyclist, or a pedestrian be able to negotiate a location like this?

Recommendation 3: Determine Lane and Shoulder Width

The following minimum values should be considered:

US Customary

Design Year ADT[a]	Speed[b]	<10% Trucks/ Machinery[c]		>10% Trucks/ Machinery[c]	
	(Mph)	Lane[e] Width	Shoulder[d] Width	Lane[e] Width	Shoulder[d] Width
1 – 750	≤45	9 ft	2 ft	10 ft	2 ft
751 – 2000	≤45	10 ft	2 ft	10 ft	2 ft
2000 >	≤45	11 ft	3 ft	12 ft	3 ft

[a] Design Year ADT should be based on a 10-year projection
[b] Speed should be based on average speed
[c] Some types of vehicles may require additional roadway widths.
[d] Roadways having curbing may have 1.5 ft width of shoulder
[e] In context sensitive environments, engineering judgment should considered the existing lane widths to remain, evaluating site-specific crash data and the possible use of traffic control devices as an alternative.

Design Year ADT [a]	Speed [b]	< 10% Trucks/ Machinery [c]		> 10% Trucks/ Machinery [c]	
	(Mph)	Lane [e] Width	Shoulder Width	Lane [e] Width	Shoulder Width
1 – 750	>45	10 ft	2 ft	10 ft	2 ft
751 – 2000	>45	10 ft	3 ft	11 ft	3 ft
2000 >	>45	11 ft	4 ft	12 ft	4 ft

[a] Design Year ADT should be based on a 10-year projection

[b] Speed should be based on average speed

[c] Some types of vehicles may require additional roadway widths.

[e] In context sensitive environments, engineering judgment should considered the existing lane widths to remain, evaluating site-specific crash data and the possible use of traffic control devices as an alternative.

Recommendation 4: Determine Normal Pavement Crown

The designer should develop consistent procedures for evaluating the existing pavement crown, with the following objectives:

- The pavement overlay should match new construction normal crown policies. Typically 2 - 2.5 % cross slope.
- The shoulder cross slope should allow rainfall to drain the roadway. Typically 4 - 6 % cross slope.

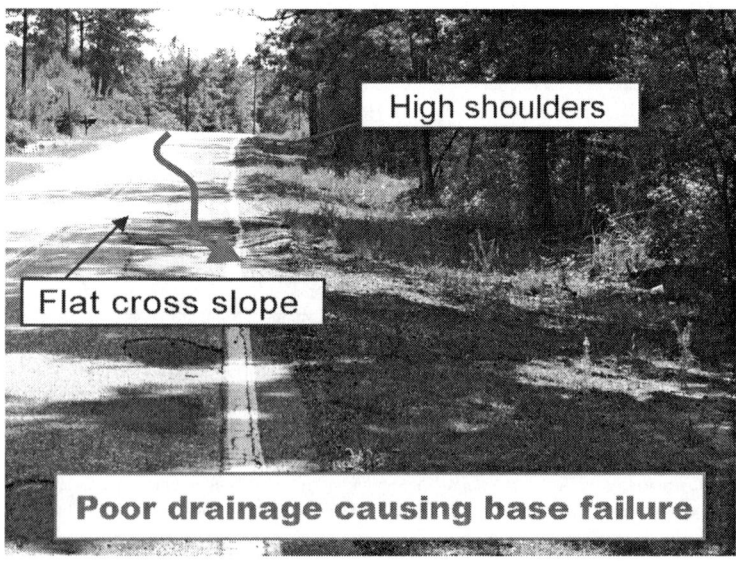

- The combination of grass shoulders, higher than the travelway, directs water down the travelway to a flat cross-slope area. The poor condition of the pavement (cracks) allow this accumulated water to percolate into the base of the roadbed causing severe damage, which is costly to repair.

Recommendation 5: Determine Horizontal Curvature and Superelevation

The designer should review each horizontal curve to determine the appropriate action that may be required. Refer to AASHTO, *A Policy on Geometric Design of Highways and Streets*, 2004 (9) for the suitable superelevation (Method 2) that should be considered. Use of a Ball-Bank indicator and its procedures is an additional tool in determining the comfort level of the vehicle based on different speeds around the curve.

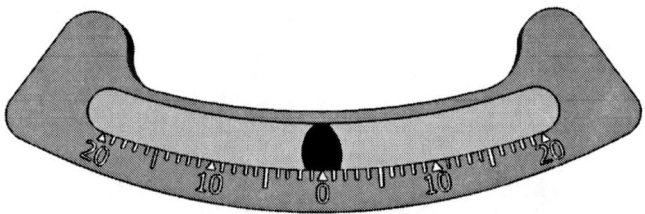

There are various types of ball-bank indicators available. When mounting this device in a vehicle it is very important to have the vehicle on a level surface.

- The designer should adjust the existing cross section with increased superelevation to match the average speed of vehicles.

- Simplified curve formula: $R_{min} = \dfrac{V^2}{15\,(0.01\,e_{max} + f_{max})}$

- It is acceptable for the designer, when evaluating curves with low average vehicle speeds, **<45 mph,** to resurface without changing the existing curve geometry and cross section if the nominal design speed of the curve is within **15 mph** of the average vehicle speeds, and if there is no clear evidence of a site-specific safety problem associated with the curve. Note: A speed study will aid in understanding the prevailing speeds and common characteristics of the users.
- The designer, when evaluating curves with high average vehicle speeds, **45 mph and higher,** should consider reconstruction when the nominal design speed of the existing curve is more than **15 mph** below the average vehicle speeds, and the projected traffic volume is greater than 1000 ADT,

or if there is a site-specific safety problem associated with the curve.

- If curve reconstruction is not feasible, additional measures should be considered to aid the driver. To reduce speed; supplemental signing, pavement markings, rumble strips, or other traffic control devices should be applied. To improve the roadside: clearing slopes, flattening steep side-slopes, or removing, relocating, or shielding obstacles, should be evaluated if there is an appreciable site-specific safety problem. To improve the roadway: widening lane width, widening shoulder width, or paving shoulders may improve the driving operation.

Procedure for the use of the Ball-Bank Indicator to determine the safe speed of a curve

The ball-bank indicator is used to measure the overturning force (side friction), measured in degrees, on a vehicle negotiating a horizontal curve. The ball-bank can be easily mounted to the dashboard by means of rubber suction cups or other stable methods. It should be mounted in such a position as to allow the ball to rest freely at the zero degree position when the vehicle is standing level. The movement of a car around a curve to the left, for example, causes the ball to swing to the right of the zero degree position. The faster the car moves around the curve, or the sharper the curve, the greater distance the ball swings away from the zero degree position. Superelevation, however, tends to bring the ball back to the zero position. The net result is the indicator reading in degrees of deflection.

Beginning well in advance of the curve being checked, the driver should enter the curve at a predetermined speed, drive the car parallel with the centerline of that travel lane, and maintain that uniform speed throughout the curve. The curve should be driven a number of times until at least two identical ball-bank readings (degrees) for each direction of travel are obtained. Each direction of travel should be considered separately.

The maximum negotiable safe speed for the curve is the speed at which the ball-bank indicator's reading is 10 degrees or less for 35 mph or greater. The first trial run is made at a speed somewhat below the anticipated maximum safe speed. Subsequent trial runs are conducted at 5 mph speed increments. Readings of 14 degrees for speeds of 20 mph or less, 12 degrees for speeds of 25 mph through 30 mph and 10 degrees for speeds of 35 mph through 50 mph are the usually accepted limits beyond which riding discomfort will be excessive and loss of vehicle control may occur.

The recommended advisory speed should be to the nearest 5 mph less than the maximum negotiable safe speed determined separately for each direction of travel. Considerations of sign distance, intersections, crash records, and other conditions may result in a recommended speed lower than that derived by the ball-bank indicator method.

Advisory speed plates should be used in conjunction with curve and turn signs when the safe operating speed is below the posted or prevailing speed on the roadway. When plates are used with curve and turn signs, the miles-per-hour value shown on each plate should be determined by the use of the ball-bank indicator. The lowest speed (to the nearest 5 mph) obtained during trial runs that creates a reading of 10 degrees or more on the ball-bank indicator shall be used (degrees and mph are stated above). Each direction should be checked independently and may be posted with different speeds.

The Manual on Uniform Traffic Control Devices, MUTCD, recommends the use of a Turn (W1-1) sign for a location where test runs at 30 mph or less has been determined for the curve. The use of a Curve (W1-2) sign is recommended for a location where test runs at speeds greater than 30 mph has been determined for the curve.

Recommendation 6: Determine Vertical Curvature and Stopping Sight Distance

The designer should review each vertical curve to determine the appropriate action that may be required.

- It is acceptable for the designer, when evaluating curves with low average vehicle speeds, **<45 mph,** to resurface without changing the existing curve geometry if the nominal design speed of the curve is within **20 mph** of the average vehicle speeds, and if there is no clear evidence of a site-specific safety problem associated with the curve.
- The designer, when evaluating curves with high average vehicle speeds, **45 mph and higher,** should consider reconstruction when the design speed of the existing curve is more than **20 mph** below the average vehicle speeds, and the projected traffic volume is greater than 1000 ADT, or there is a site-specific safety problem associated with the curve.
- If curve reconstruction is not feasible, additional measures should be considered to aid the driver. To reduce speed; signing or other traffic control devices should be applied. To improve the roadside; removing, relocating, or shielding location of driveways or intersecting roads should be evaluated if there is an appreciable site-specific safety problem. To improve the roadway: lengthening sharp horizontal curves, widening a narrow bridge, or improving other geometric features adjoining the vertical curve proximity may improve the driving operation.
- Sag vertical curves typically do not create sight restrictions and do not have to be reconstructed, unless there is a site-specific safety problem.

Recommendation 7: Determine Bridge Width

The designer should evaluate bridge replacement or widening if the bridge is less than 100 ft. long and the usable width of the bridge is less than:

US Customary

Design Year ADT [a]	Speed (Mph)	Usable Bridge Width [b, c, d]
1 - 1000	All Speeds	Width of approach lanes
1001 – 4000	≤ 45	Width of approach lanes plus 2 ft
1001 - 4000	>45	Width of approach lanes plus 3 ft
4000 >	≤ 45	Width of approach lanes plus 3 ft
4000 >	>45	Width of approach lanes plus 4 ft

[a] Design Year ADT should be based on a 10-year projection

[b] If the roadway width (lane plus shoulder) is paved, the bridge should be equal in width

[c] Bridge usage by trucks, farm machinery, or recreational vehicles should be considered in determining the appropriate width

[d] Existing bridges may remain in place without widening unless there is evidence of a site-specific safety problem

Is one 12' section of guardrail enough to help in a crash? How about the end-treatment?

8" drop off

- If bridge replacement is not feasible, the designer should evaluate the approaches to the bridge and to implement additional measures that may aid the driver. Installing transition guardrails, advance warning signs, and/or other traffic control devices should be considered.

Recommendation 8: Determine Side Slopes and Clear Zones

The designer should develop consistent procedures for evaluating and improving roadside features with the following objectives:

- A clear zone of any width should provide some contribution to safety. Thus, where clear zones can be provided at little or no additional cost, their incorporation in design should be considered. A 2 - 3 ft. shoulder is recommended for speeds ≤ 45, and 2 – 4 ft for speeds greater than 45.
- Retain current slopes (without increasing front slopes) when widening lane and shoulders, unless warranted by special circumstances.
- Flatten side slopes steeper than 3:1 at site-specific locations where there is evidence of a crash or available crash data.
- Remove, relocate, or shield isolated roadside obstacles.
- Crossdrain pipes and culverts should only be extended as required to provide the width for the pavement, shoulder, and conform to the existing side-slope where possible. Headwalls may be retained on existing crossdrain structures where there are no adjustments required for the pavement and shoulder widths. Site-specific crash locations should be evaluated.
- Sidedrain pipe should be relocated as required to obtain the width for the pavement, shoulder, and to match existing side-slopes along the roadway as possible. Slope-paved headwalls of other sloped-end treatments should be provided. Headwalls may not be replaced on existing sidedrain pipe that will remain in place if no adjustments are required for the pavement or shoulder widths. Consideration should be given to replacing large vertical headwalls that are close to the pavement and are a potential hazard. Site-specific crash locations should be evaluated.
- When it is not feasible to make improvements to the clear zone, because of: terrain, right-of-way, potential social / environmental impacts and/or cost, the provision for a clear recovery area may be impractical to achieve. Clear recovery areas of a width that is less than desired may be used. Engineering judgment should be used to implement the use of traffic control devices, if warranted, to assist and warn the driver where there may be an appreciable site-specific safety problem.

- Do the headwalls need to be this tall? If so, should a reflector of some type be installed to indicate its presence? Should the pipe under the driveway be this size? A drainage study may determine a smaller diameter.

- Does the driver have any idea that a major highway is at the top of the incline? Just how difficult would it be at night? A local driver may know this, but a first time user?

Recommendation 9: Guardrail Need for Embankments and Culverts

The designer should develop consistent procedures for evaluating the need for guardrail, with the following considerations:
- Examining the shoulder slopes and culvert sizes.
- Identifying site-specific safety locations.
- Clear zone encroachments

The following charts are guidelines from Transportation Research Record 1599, *"Guardrail Need: Embankments and Culverts"*, Transportation Research Board, Washington, DC, 1997 (6) and is intended to be used as tools to aid the designer in the decision making process. These curves are intended to eliminate the need for conducting benefit-cost analysis. These charts may be used if the slope or culvert is within the clear zone, or if there is a site-specific safety problem.

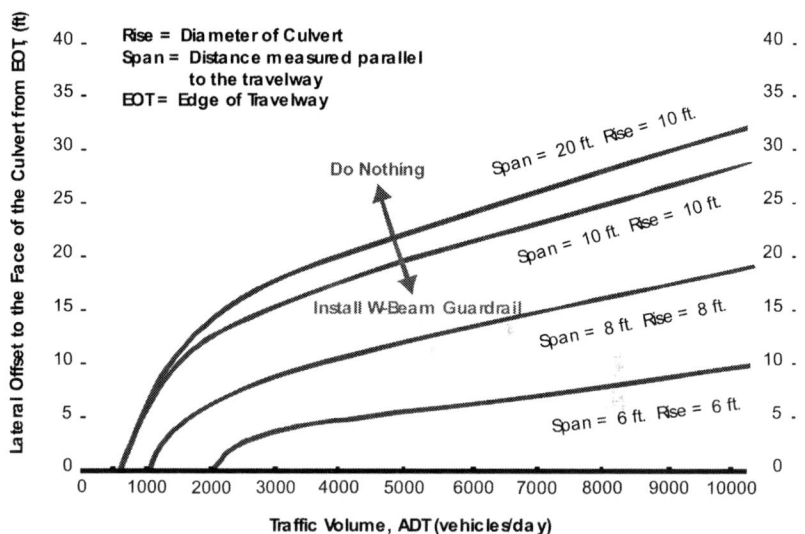

W-Beam Guardrail Need for Culverts

Source: Guardrail Need, Embankments and Culverts, Transportation Research Record 1599, 1997

The previous chart illustrates the lateral offset from the travelway to the face of the culvert. It shows the correlation between various ADT (vehicles per day) volumes and the various culvert sizes, depth (rise) of the culvert and its minimum length (span) along the travelway. If the culvert being evaluated falls below the various curved lines shown in the chart, a guardrail should be considered to be installed.

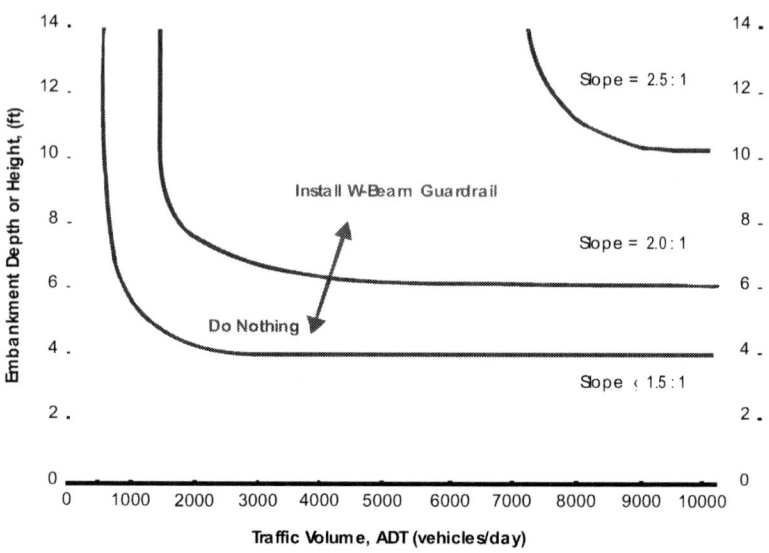

W-Beam Guardrail Need for Embankments

Source: Guardrail Need, Embankments and Culverts, Transportation Research Record 1599, 1997

This chart illustrates the embankment depth from the outer edge of the roadway shoulder (elevation), down various slope rates (frontslope), to the lower elevation of the adjacent terrain. It shows the correlation between various ADT (vehicles per day) volumes and the various depths based on the frontslope rates. If the depth being evaluated falls above the various curved lines shown in the chart, a guardrail should be considered to be installed.

Recommendation 10: Pavement Edge Drop and Shoulder Type

The designer should develop consistent procedures for evaluating pavement edge drop conditions and the type of shoulder construction, with the following objective:

- All shoulders should be re-established and graded to a consistent slope.
- Edge of pavement drops should be repaired and should match the shoulder slope.
- Selectively pave shoulders at points where there are site-specific safety problems (outside or inside of horizontal curves, across from intersecting roads, etc.).

Recommendation 11: Intersection Improvements

The designer should develop consistent procedures for evaluating intersection improvements, with the following:

- Collision diagrams showing vehicle paths, time of occurrence, and weather conditions.
- Condition diagrams showing important physical features that affect traffic movements.
- Field review of the intersection to detect hazards not apparent from collision and condition diagrams.
- Designer should consider intersection improvements to site-specific safety problem areas.
- Improvements may be organized on three primary design objectives: reduction of potential conflicts (traffic signals, turn lanes, etc.), improves driver decision-making (longer lines of sight, lane markings, etc.), and improves the braking capability of the vehicle (warning signs, increased pavement skid resistance, etc.).

Recommendation 12: Document the Design Process

Before developing construction plans and specifications, designers should prepare a safety and design report based the above 11 recommendations. Additional information regarding specific elements, not mentioned above, may be included in this report.

For some RRR projects, it may be necessary to have this document submitted to an appropriate agency that is responsible for the project area for review and approval. The format of the project file will be established by the same agency.

Any waivers of the design criteria shall be submitted to and approved by the same agency or their governing authority having project approval. It is understood that design waivers may not be needed for RRR projects if the project is internal to the same agency funding the project. However, fully documented project information should be compiled and filed.

References

1.   Transportation Research Board, *"Special Report 214"*, 1987.
2.   AASHTO, American Association of State Highway and Transportation Officials, *"Guidelines for Geometric Design of Very Low-Volume Local Roads (ADT ≤ 400)"*, 2001.
3.   AASHTO, Association of State Highway and Transportation Officials *"Geometric Design for Resurfacing, Restoration, and Rehabilitation (RRR) of Streets"*, 1977.
4.   AASHTO, Association of State Highway and Transportation Officials, *"Roadside Design Guide"*, 2002.
5.   National Cooperative Highway Research Program, *"Report 350, Recommended Procedures for the Safety Performance Evaluation of Highway Features"*, 1993.
6.   Wolford, D., and D.L. Sicking, *"Guardrail Need: Embankments and Culverts,"* In *Transportation Research Record 1599*, Transportation Research Board, National Research Council, Washington D.C., December, 1997. Figure 6, p. 54 and Figure 8, p. 55. Reproduced with permission of TRB.
7.   AASHTO, Association of State Highway and Transportation Officials *"A Guide for Achieving Flexibility in Highway Design"*, 2004.
8.   AASHTO, Association of State Highway and Transportation Officials , *"Roadside Design Guide"*, 2006.
9.   AASHTO, American Association of State Highway and Transportation Officials, *A Policy on Geometric Design of Highways and Streets*, 2004.